9/18

WITHDRAWN

Hassenfeld Library
University School of Nashville
2000 Edgehill Avenue
Nashville, TN 37212
www.usn.org

HANDS-ON ROBOTICS™

THE FUTURE OF ROBOTICS

Hassenfeld Library
University School of Nashville
2000 Edgehill Avenue
Nashville, TN 37212
www.usn.org

LAURA LA BELLA

New York

Published in 2018 by The Rosen Publishing Group, Inc.
29 East 21st Street, New York, NY 10010

Copyright © 2018 by The Rosen Publishing Group, Inc.

First Edition

All rights reserved. No part of this book may be reproduced in any form without permission in writing from the publisher, except by a reviewer.

Library of Congress Cataloging-in-Publication Data

Names: La Bella, Laura, author.
Title: The future of robotics / Laura La Bella.
Description: New York : Rosen Publishing, 2018 | Series: Hands-on robotics | Includes bibliographical references and index. | Audience: Grades 5–8.
Identifiers: LCCN 2017024559| ISBN 9781499438901 (library bound) | ISBN 9781499438888 (pbk.) | ISBN 9781499438895 (6 pack)
Subjects: LCSH: Robotics—Juvenile literature. | Robots—Juvenile literature.
Classification: LCC TJ211.2 .L3 2018 | DDC 629.8/92—dc23
LC record available at https://lccn.loc.gov/2017024559

Manufactured in China

CONTENTS

INTRODUCTION 4

CHAPTER ONE
RISE OF THE ROBOTS 7

CHAPTER TWO
WHAT ROBOTS CAN DO 13

CHAPTER THREE
A WORLD OF ROBOTICS 21

CHAPTER FOUR
THE DOWNSIDES OF THE COMING ROBOT ERA ... 26

CHAPTER FIVE
YOUR FUTURE IN ROBOTICS 32

GLOSSARY 39
FOR MORE INFORMATION 40
FOR FURTHER READING 42
BIBLIOGRAPHY 43
INDEX 46

INTRODUCTION

Everywhere you look, you can find robots engaged in activities both simple and complex. Robotics combines computer science, engineering, and electronics to create devices that can perform a variety of tasks, many of which are too complicated, too precise, or nearly impossible for humans to perform. What defines a robot from merely a machine that can accomplish a task is advanced computing.

Robots are programmed to make limited decisions based on certain scenarios. That's why a robotic arm on a manufacturing floor can accurately weld vehicle parts together or install microchips on a circuit board, and recognize if something is not quite right in the process. This simple version of artificial intelligence (AI) enables the robot to spot mistakes in manufacturing as they occur or are about to.

Right now, robotics is being utilized in manufacturing to help assemble automobiles, furniture, and prepare and package a seemingly endless stream of other products. It is used to automate apparatuses to explore the depths of our oceans and the farthest reaches of space. In hospitals, on farms, and in our national security and defense, robotics is poised to take over many tasks in the very near future.

Robotics may also deliver highly advanced technologies on a very tiny scale. Nanobots, or tiny robots, could one day swim through our bloodstreams to deliver medications to specifically targeted sites within our bodies to treat a wide range of illnesses and diseases. Robofish are being tested in the waters off Spain to detect and locate contaminated water and report it back to land. And, in a fascinating partnership that may reduce the time it takes to rescue victims of natural disasters, snakebots are teaming up with rescue dogs. The dogs will sniff out

INTRODUCTION | 5

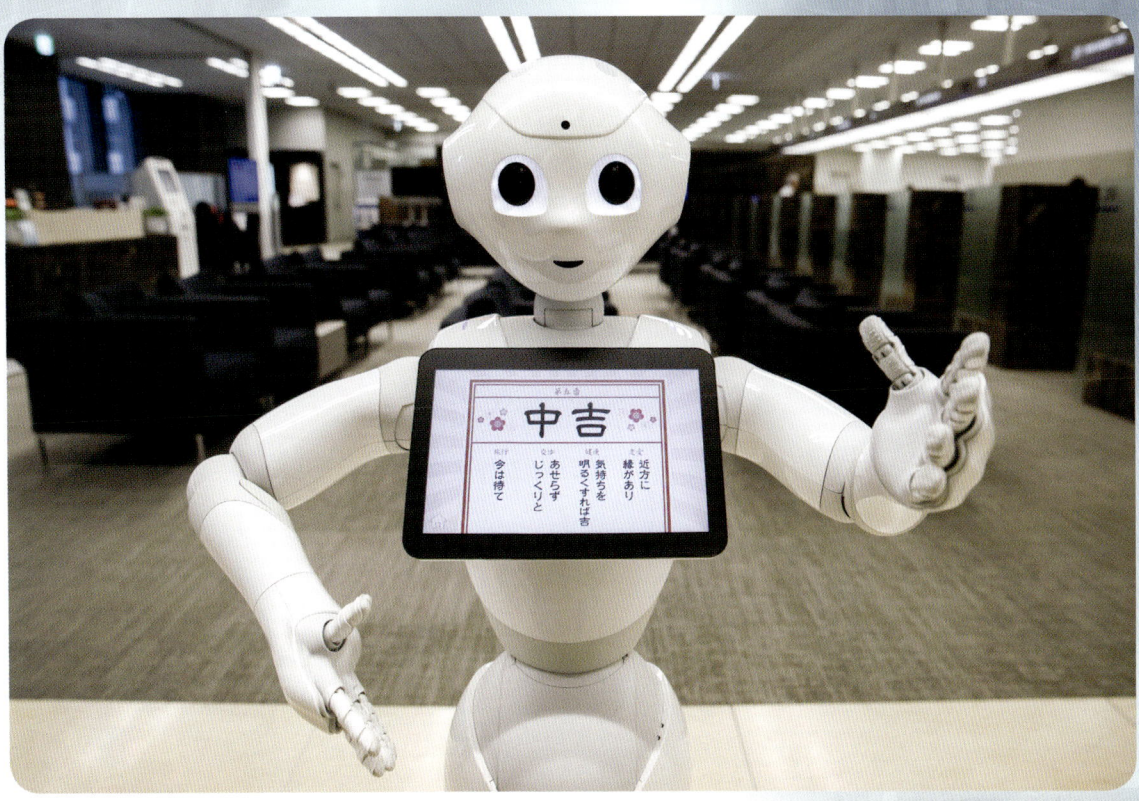

A Pepper humanoid robot, designed by SoftBank Group as a companion robot to recognize and respond to human emotion, stands inside the Mizuho Financial Group headquarters in Tokyo, Japan, in April 2017.

and locate a disaster victim and deploy a snakebot, which can relay video and audio of the situation back to medical personnel and rescue workers.

The future of robotics is exciting and promising. As new technology emerges, inspired by the innovations of scientists, engineers, and programmers, the ways society at large uses robotics will surely multiply. It is hoped that robotic technologies will improve and streamline processes in manufacturing, health care, law enforcement, education, entertainment, and more. Artificial intelligence, coupled with robotics, will create machines that are highly interactive, that anticipate our needs and act to meet them, and even be able to solve problems without our prompting them to do so.

The possibilities for the future of robotics remain great. The only limits seem to be the human imagination and how much work and financial resources we are willing to throw at them. In some ways, the future is not far off. It's now.

CHAPTER ONE

RISE OF THE ROBOTS

Henry Ford, the pioneering automobile manufacturer, first automated the assembly line in car production in 1913 with a conveyor belt. The belt had been used since the late eighteenth century in mines and rail yards to move heavy material and equipment, but now it was put to use to expand automobile production. Ford had machines built that could stamp out car parts automatically and with more precision and speed than his human workers. All these measures vastly improved the rate of production. Ford's innovations reduced the time it took to manufacture a car from twelve hours to just under three. It was the first automated assembly line and the first time basic robotics were used in manufacturing. Manufacturing, and all the various industries that would soon employ it in mass production, would never be the same.

WHAT ARE ROBOTS AND HOW DO THEY WORK?

When many people think of robots, they might imagine robots that look like humans and that are capable of communication and accomplishing the types of tasks that people normally perform. For most of the modern era, however, this kind of robot has actually been quite uncommon and has been restricted mostly to science fiction, including novels, stories, films, and television.

While humanoid robots exist, it is far more common to encounter a robot that embodies a part of a human shape, such as a robotic arm, or a machine that incorporates robotic technology but looks more machine than human. Most leading roboticists—the specialists who develop, design, and

This image from around 1913 shows the automobile assembly line at the Ford Motors plant at Highland Park, Michigan. Ford's innovations in mass production included some of the successful and influential examples of automation.

build robots—have defined a robot as having a programmable brain, or computer, with a physical body that can move. Using this definition, robots are different from machines, such as cars or lawnmowers, because they have advanced computing that helps control their movements, rather than direct human control. Some can even make simple decisions, and these decision-making abilities are likely to increase vastly in the coming years and decades. Of course, even cars and planes themselves have begun to incorporate simple systems of artificial intelligence and robotics, and these two will grow more advanced.

Robots are mechanical devices that are capable of performing tasks on command or according to the directions they are programmed to follow. While there are many automated services that seem to resemble robots in a variety of ways—for instance, self-checkout machines at the supermarket, automated telephone systems, etc.—these systems are actually limited in their ability to interact with you. They cannot make decisions, nor can they adapt to their surroundings.

The inner workings of a bank's automated or automatic teller machine (ATM) are shown here.

Advanced robots that exist now or that are due to be rolled out soon will navigate specific problems with little or no human input. While cars are not robots, advanced modern ones can make simple, semiautonomous decisions. Cars with cruise control, self-parking abilities, collision warning systems, and dynamic headlights are all examples of simple robotic technologies.

THE MODERN EVOLUTION OF ROBOTICS

Earlier civilizations invented humanoid creations similar to robots. The ancient Greeks wrote stories about humanlike, self-operating machines called automatons, some of which appeared in their mythology. Real automatons even existed in many ancient and recent civilizations, but they were actually simply engineered human or animal-like models made to move in simple and limited ways. For example, a cuckoo clock might be thought of as a simple automaton.

It was especially during the past two centuries that some important milestones pushed forward the possibilities of making true, autonomous robots:

- **1913** Henry Ford uses the world's first automated conveyor belt on an assembly line at his factory.
- **1920** Karel Capek, a Czech writer, first uses the word "robot" to describe an artificial human being.
- **1937** Alan Turing, a famous British mathematician, writes a paper, "On Computable Numbers," which is largely credited with jumpstarting the computer revolution.
- **1950** Turing creates a test to determine if a machine has the capacity to think for itself. Known as the Turing test, it tests a machine's ability to exhibit behavior that is indistinguishable from a human during conversation.

RISE OF THE ROBOTS | 11

- **1954** Americans George Devol, inventor, and Joe Engleberger, physicist and engineer, introduce the first robotic arm. It evolves into the first industrial robot, named Unimate, which performs repetitive tasks on a General Motors assembly line.
- **1986** The first LEGO-based educational products are sold, and later, in **1998**, LEGO launches its first Robotics Inventions System, putting robotics in the hands of children everywhere.
- **1993** Dante attempts to explore Antarctica's Mount Erebus volcano. Dante is a remotely controlled robot that could collect data and explore a dangerous location inaccessible to humans.
- **1997** A computer built by IBM, named Deep Blue, competed against and won a chess game against Garry Kasparov, a world chess champion.
- **2008** The Roomba robotic vacuum cleaner sells more than 2.5 million units and launches a second phase of robotic household products.

Airline kiosks in airports that check passengers into their flights are considered robotic. They are programmed to interact with humans much like human airline representatives do and to anticipate their questions and offer solutions. These computing systems are designed to monitor and respond to scenarios in real time and as scenarios change. A combination of engineering and computer programming are essential for a robot to work.

ANATOMY OF A ROBOT

Many robots are designed to act as stand-ins for humans. They can work in locations that are relatively inaccessible and tough (or impossible) for humans to inhabit or visit, such as remote parts of the ocean or in outer space. Robots' precision and extra-human abilities can also allow them to perform tasks humans cannot. For example, robots can mass produce the tiny circuit boards used in smartphones.

The three main parts of robots are based on those of the human anatomy, even for those robots that are not quite humanoid in appearance: a body, a nervous system, and a brain.

- **The body**. This portion of a robot is usually created by an engineer, who focuses on the physical aspects of the robot's intended movements. It executes the intended motions and tasks.
- **The nervous system**. This central communication hub includes the electronics, embedded systems, and computer programming used to communicate messages between the brain and the body of the robot. Electrical and electronic engineers are the ones who make sure the brain's message to drill a hole in sheet metal is understood and executed properly by the arm of the robot.
- **The brain**. As with a human, the brain of a robot generates and sends messages that determine the movements and other actions of the body via its nervous system. Roboticists that design and build a robotic brain might be programmers in computer science or work in related fields such as AI, electrical engineering, human-computer interaction, or advanced computing.

CHAPTER TWO

WHAT ROBOTS CAN DO

While humanoid robots exist, most robots don't resemble people. They come in all shapes and sizes, and may soon be found throughout society. Even areas where robots do not have a foothold yet are on the verge of incorporating them—especially in the household realm.

MANUFACTURING AND INDUSTRY

Automation refers to engineering, computing, and electronic systems that operate automatically with little to no human direction, and it has altered automobile and high-tech manufacturing. These technologies are extremely versatile and can be found in food production and processing, where robots control the measurement of ingredients, serving sizes, and packaging. Most products one may encounter daily probably was developed at least partially with some kind of automation. Even after a product is completed, robots might package it and deploy it to different destinations, including onto trucks, trains, or air freight carriers.

The introduction of robots has improved quality, speed, and accuracy in production, while decreasing production costs. They are able to work with a speed and efficiency that humans cannot match, and they have the ability to roll out identical versions of the same thing repeatedly with incredible accuracy.

Used primarily in industrial settings, industrial robots are programmed to accomplish specific tasks: repeatedly punch out a metal car part, sew upholstery, or install an engine. They can also move or haul heavy materials and parts. Monitored by humans, they complete tasks with little human interaction. These robots can be extremely large and heavy, especially those used in heavy manufacturing. Their activity can also be hazardous to humans. Thus, they often work behind protective fencing or reinforced plastic or glass, which also shields human operators from sparks or heat.

One example of a robot in modern manufacturing is this one moving a glass panel through the Hevel Solar factory, which builds solar energy modules in Novocheboksarsk, Russia.

Other kinds of industrial production also employ robotic elements. Dangerous chemicals and biotechnological organisms that can be deadly to humans can be made in facilities largely using automated robotics. The manufacture of any products that requires extreme heat or cold to maintain safety or simply the integrity of the products themselves is a natural fit for automation.

SERVICE AUTOMATION

The service industry, made up of businesses that cater to customers and clients, is experiencing an increased robot presence. Robots are on the verge of replacing many customer service jobs. Many fast foot outlets worldwide have rolled out touchscreen-based, automated ordering systems. Future franchises may have friendly, humanlike robots that take orders, prepare them, and handle point of sale (POS) systems.

Robots might replace humans working in retail. A department store drone might greet you, offer to take your clothing size, and even incorporate other information the store previously gathered on you (previous purchases there, for example) based on photographic recognition, and then offer you clothes you might want to try on or buy.

Hotels and other hospitality businesses could also replace humans who clean and maintain rooms and common areas, wash and fold linens, take room service orders, and check customers in and out.

DOMESTIC ROBOTS

The Neato Botvac, an automated vacuum robot, will recognize when its battery is low and automatically return to its charging station. The Grillbot cleans the grates on a barbecue grill. The Litter-Robot takes care of cleaning a cat's litter box, while the Robomow RS mows lawns. Domestic robots are designed to make life easier for those who are too busy or find household chores too overwhelming or tedious.

As for future projects, the old image of a humanoid robot housekeeper is perhaps not very far off the mark. On the other hand, some may simply be dronelike robots that move along floors or even travel along walls and ceilings to clean, among other tasks. Homes may be equipped with artificial intelligence

Attendees of the 2015 World Robot Conference in Beijing, China, marvel at a cleaning robot moving vertically up a glass barrier. There will probably be few surfaces future robots won't be able to clean.

that networks and coordinates work among various robots and other household systems (like cooling, heating, water, sewage, and laundry functions), and times these processes to occur while the home's residents are out or even asleep.

MEDICAL ROBOTS

Robots used in hospital and pharmacy settings are becoming increasingly common. Surgical robots assist physicians in operating rooms, allowing surgeons to perform delicate procedures with great accuracy and increased safety. Modern robotic surgery is controlled largely by hand by human operators, but future ones will almost certainly be far more automatic and intuitive.

Tugs are robots manufactured by Aethon, a company specializing in delivery automation. They are autonomous, mobile robots that move around a hospital delivering medicine, lab results, and supplies, and also collect medical waste and garbage. They are designed to navigate a hospital and avoid collision with staff, patients, and visitors.

The da Vinci surgical robot could change how surgeons operate. The robot's multiple arms can be fitted with instruments like scalpels or scissors, and a separate console system is controlled by a surgeon. Doctors can perform complex procedures precisely and that are less invasive and debilitating to patients. Robots like da Vinci could revolutionize medicine. One innovative change will be the ability of both human-operated and preprogrammed machines to perform operations remotely, even via the internet from thousands of miles away—a movement some call "telemedicine." Such developments could help bring doctors to impoverished regions of the world and allow for greater flexibility for patients and doctors alike.

ROBOTS IN POLICE WORK AND AT WAR

Law enforcement and military operations can be dangerous. There are places soldiers and police officers would rather avoid going or entering. Enter robots, which can perform surveillance, reconnaissance, and explosives detection and removal without endangering police or troops. For instance, the Cleveland, Ohio, police force has enlisted a new recruit, a robot named Griffin. According to *Wired*, the six-wheeled robot is about 12 inches (30.5 cm) tall and can go places officers cannot, including unsafe and hostile environments. Griffin allows users to monitor or investigate crime scenes or active situations from safe distances.

Large tactical robots often require trucks to deliver them to a scene. Short for Bomb Assault Tactical Control Assessment Tool, the BatCat is used by the Los Angeles Police Department to remove large explosive devices or to literally crash through walls if necessary. It has a 50-foot (15 m) telescoping arm that can handle explosive devices from afar.

The military regularly uses robots in a variety of ways, including to disarm improvised explosive devices (IEDs) and to conduct reconnaissance and surveillance. Future military robots will include more machines that will actively engage in combat. The US military is also building robots designed to retrieve

This overhead shot shows a New York Police Department robot at work after disposing of a bomb following an explosion that injured twenty-nine people on September 17, 2016.

injured personnel from the battlefield. Automating rescue may take battlefield medics out of harm's way. Robots could also perform surgery or administer first aid on the spot, even under fire, or deliver needed medical supplies.

Jason Healey, a senior fellow on the Atlantic Council's Cyber Statecraft Initiative, told *The Hill* that he expects robot soldiers to become operational within the coming decade. Human combatants may also be augmented with robot parts to increase strength, accuracy, and stamina. The stuff of science fiction may

SUPERSOLDIERS

Research is currently under way at the Massachusetts Institute of Technology's (MIT) Institute for Soldier Nanotechnologies, which is working with the US military and various industrial partners to develop and test robotics and nanorobotics that could enhance human abilities in as yet unseen but very dramatic ways. For example, the term "supersoldiers" is being used to describe soldiers that could be enhanced by robotics to increase their performance and even self-administer medical aid. The main tools for achieving this goal are nanorobots, tiny microscopic robots that could be injected into the body to release medications and drugs, or help speed healing. Nanorobots could help soldiers recover quickly from injuries or the wear and tear caused by things like long marches, extreme temperatures, or sleep deprivation.

become reality if robotic limbs and other organ systems can help military amputees and the injured regain full abilities.

SCIENTIFIC ROBOTS

The National Aeronautics and Space Administration (NASA), the United States' federal space agency, has made use of robotics for some time. Space shuttles have the Canadarm to assist in operational missions. The Mars Rover can navigate another world to collect rock and soil samples. Without the Mars Rover, scientists may never have learned that Mars has water beneath its surface.

Scientific robots also conduct experiments and complete repetitive tasks in research, freeing up scientists to work on their core missions. They can also conduct search efforts in deep water to seek out shipwrecks or planes that have gone missing. Robotics will likely become more prevalent in all aspects of space exploration, resource extraction, and many other tasks that the future will require.

CHAPTER THREE

A WORLD OF ROBOTICS

A crawling, cylinder-shaped robot tasked with gathering information about a particular geographic area suddenly shifts its shape into a buzzing helicopter and takes flight to gain a bird's eye view of the terrain. It might sound like something from the blockbuster *Transformers* action film franchise, but it is really a prototype being developed by researchers associated with the Center for Distributed Robotics at the University of Minnesota. This transforming robot is helping to improve mapping and topography techniques by combining data from both ground-based and airborne robots. This is but one example of a major change that robotics will make possible in coming decades.

A customer interacts with a robot server at a Shenyang, China, restaurant in March 2016. Humanoid service robots could very well be a dominant element within a few decades.

21

A FUTURISTIC WORLD LIES AHEAD

A robotic moon base, self-driving cars, and robot miners digging on distant asteroids once seemed like the stuff of science fiction, but they may very well become reality in the coming decades with the vast progress expected in robotics and engineering.

Japan, for example, has future plans for a robotic moon base that will be staffed by androids possessing advanced artificial intelligence, are self-repairing when they break down, can multi-task, and can perform scientific exploration on the moon.

A handful of auto manufacturers are developing prototypes of advanced, self-driving cars. The end products will likely have sophisticated mechanisms for communicating with other motorists and automated vehicles, and be able to anticipate and avoid collisions and other traffic hazards. These future machines may be plugged into complex networks that may help people manage highways and roads super efficiently.

Space travel to Mars and to other faraway destinations would put untold pressures on the human body. Research is underway to determine how humans may respond to longterm travel in our quest to understand other planets as well as the outer reaches of our solar

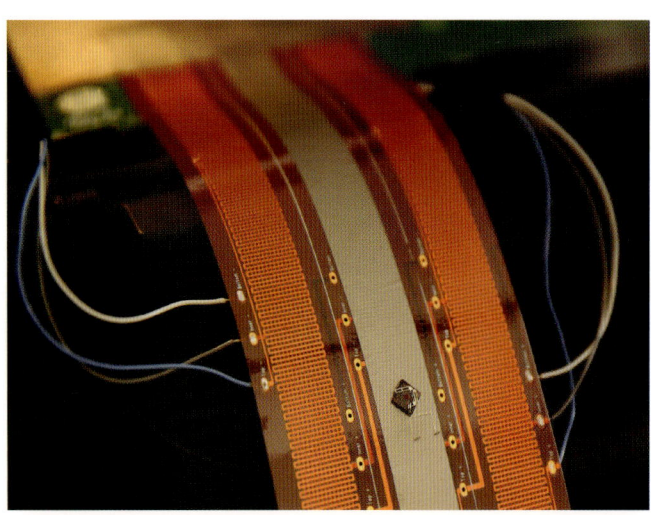

The tiny object located in the middle of this circuit ribbon is actually a microrobot, and it is one of the tiny machines featured at the Defense Advanced Research Projects Agency (DARPA) Robotics Challenge Expo in Pomona, California.

A WORLD OF ROBOTICS | 23

An employee demonstrates the push-button technology of a Samsung Electronics Family Hub refrigerator in April 2016 at the recently launched Smart Home section of London, England's, John Lewis department store.

system. Deploying robots in roles normally held by astronauts and scientists may buy the space program time, while we learn if the human body can handle prolonged space travel.

ROBOTS AS "EMPLOYEES"

Companies and organizations recognize the potential of robots in the workplace. There are many advantages, though most of them benefit company owners, upper management, stockholders, and consumers of products and services much more than workers, whose families and communities also depend on the wages and benefits that automation will wipe out. Some include:

- **Increased productivity**. Robots can work longer shifts that human workers, and at a higher speed than their human counterparts. This means more products can be produced in a shorter amount of time.
- **Improved quality and reliability**. A robot follows its pro-

SMART FACTORIES WILL CHANGE MANUFACTURING

Smart factories, or what's being called Industry 4.0, are the integration of our physical and digital worlds. Engineers think of them as interconnected webs of information and production where robots are being fitted with sensors, software, and machine-to-machine learning that allow them to collect data as they work to help engineers to analyze data that can impact the productivity of each robot. Robots will both provide labor, and collect information and data automatically as they engage with their immediate environments. Engineers can then analyze this information to improve efficiency and productivity and decrease time and costs. The on-site managers will more closely resemble information technology specialists one might find at an office than human personnel supervisors, foremen, or other current staffers in modern industrial settings.

grammed instructions. So each part is manufactured with accuracy and precision, and meets the prescribed specifications. Parts are likely to be of uniform quality.
- **Reduced waste**. Robots have such high accuracy that they utilize materials to their fullest, producing less waste than humans, who are more prone to making errors and using excess materials. Human error may also result in greater losses in terms of faulty or lower-quality products than desired.
- **Safer workplaces**. Robots can carry heavy loads, work with hazardous materials, and perform well in environments that are dangerous for humans.
- **Less Hassle**. Some suggest that sometimes messy interactions between humans—including arguments, resentment among workers or toward superiors, sexual harassment, bullying, and numerous other issues—can be completely avoided with a robotic workforce, at least in some types of workplaces. In addition, robots don't show up late, take lunch or breaks, call in sick, or go on vacation. They can work without rest or sleep.
- **Cutting Costs**. After initial investment, robots do not require pay, retirement benefits, or health care expenditures.
- **Other Financial Savings**. Robots work fast and accurately, which means products are produced in less time and with few, if any, errors, getting products on the market faster.

CHAPTER FOUR

THE DOWNSIDES OF THE COMING ROBOT ERA

Will humans thrive in a world where robots become ever more numerous? Will the robotic revolution take over completely, displacing workers until only highly skilled jobs are left for the ultraeducated? These are important things to consider as more and more of our world comes to rely on robots and robotics.

A study by the Pew Research Internet Project and Elon University's Imagining the Internet Center asked more than 1,800 technology experts about the future of man vs. robot in the year 2025. Fifty-two percent thought that the potential robotics has for industry and the workforce has the capacity to create new jobs. A little less than half, or 48 percent, thought otherwise. This group predicted that robots would take more jobs away from American workers, leaving the working class with shrinking options.

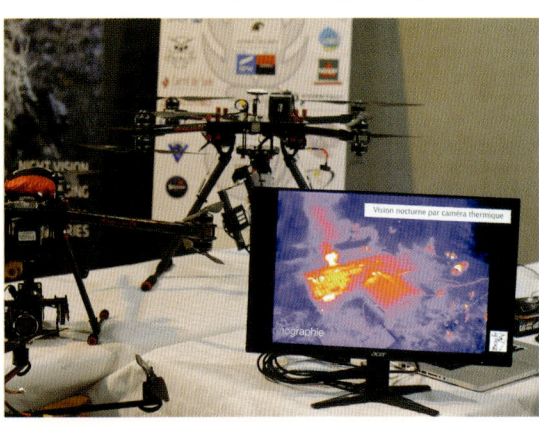

These drones and their control screens are displayed at a police barracks in Lyon, France. Law enforcement's use of drones raises concerns about privacy.

WORKFORCE DISRUPTIONS

According to an article in the *New York Times*, robots are the cause for more than 670,000 manufacturing jobs between 1990 and 2007 to be no longer performed by humans. Job losses in these sectors will continue to rise as innovations create robots that are ever more advanced and can thus perform more delicate and intuitive tasks.

Robots are being used in more workplaces (manufacturing and otherwise). They are both taking over some jobs from human employees and also decreasing the occupational hazards associated with many perilous jobs and industries.

In law enforcement, a robot can now search for explosive devices or for suspects in dangerous circumstances that would put officers' lives at risk. In retail, robots are being used to move heavy shipment of clothing or consumer products, stock shelves, fold clothing, and clean stores. In farming, robots are harvesting crops, pruning vegetation, spraying pesticides, and monitoring the growth of crops and livestock.

Engineers and scientists predict an initial return of some manufacturing jobs and the need for highly skilled workers. First, the rollout of robots is expected to bring some manufacturing jobs back to the United States. These will likely include those who manufacture robots, even alongside robots on production and assembly lines. However, the eventual and unfortunate result will be that an ever wider cross-section of the working class will be displaced by robots.

Second, this increased use of automation is expected to create a demand for qualified manufacturing engineers and technicians who will be needed to help develop and produce robots, perform maintenance on these advanced machines, and monitor and improve robot-based manufacturing processes.

THE FUTURE OF ROBOTICS

While working class positions will decrease sharply, there will be an increased need for workers with advanced technical skills. According to the US Bureau of Labor Statistics, there will be an annual need of between fifty thousand and one hundred thousand people to fulfill jobs in advanced manufacturing each year, and that number is expected to grow as more companies find increased uses for robotics. These jobs will usually require college degrees and often graduate degrees in engineering, computer science, and related fields

Fast food franchises rolling out automated ordering, like this one in Shanghai, China, may one day be the rule rather than the exception.

THE LIMITS OF ROBOTICS

Regardless how many jobs robots take over now or in the future, there are vulnerabilities in automating large chunks of the workforce. When nearly every industry is interconnected digitally, viruses and hacking becomes a serious issue. If our national energy grid was entirely online, could hostile parties hack into it and plunge the United States into darkness? Can skilled hackers gain control of robots at a factory and wreck havoc or hurt human workers?

In addition, if homeowners can check on their pets using smart home technology, what prevents thieves from gaining access to our homes through these very same systems? Conveniences incorporated to help manage home life more efficiently may also leave people vulnerable to theft, arson, and even blackmail.

Or, imagine robotic first responders. Replacing firefighters with fireproof robots that can access burning buildings might seem like a great idea. But robots do not have judgment or flexibility and are limited in acting beyond their programming. Unless artificial intelligence develops that can approximate human reasoning, we will need humans to make these judgment calls. Compassion, instinct, and other human traits are big reasons that robots will probably not take over the jobs of police, firefighters, emergency medical technicians, and similar personnel.

CAN ROBOTS TURN AGAINST US?

In the film *I, Robot*, in the year 2035 intelligent robots have been placed in public service positions around the world. But when a robotics company founder is murdered, a detective uncovers a conspiracy by robots to enslave humans. It is a far-fetched film

HAWKING'S WARNING

Stephen Hawking, the famous physicist, believes that this issue will be particularly pressing within the coming century. Hawking supports some uses of AI. He cannot speak and is paralyzed, so he uses a technical system that helps him communicate as well as conduct research. Even though he benefits from AI-related technology, he's cautious about how far humans should go in developing robots that can learn on their own.

Hawking told an audience at the Zeitgeist Conference, a group that supports global sustainability and solutions to social problems, that in one hundred years, a threat will come from machine learning. By this time, he and others believe robots will actually be able to teach themselves and make more decisions independent of human control.

The physicist also told BBC News, "The primitive forms of artificial intelligence we already have, have proved very useful. But I think the development of full artificial intelligence could spell the end of the human race. Once humans develop artificial intelligence it would take off on its own, and re-design itself at an ever-increasing rate. Humans, who are limited by slow biological evolution, couldn't compete, and would be superseded."

plot, but serious thinkers realize that a human-created, autonomous intelligence can be both a benefit and a threat.

Among the technology leaders with real concerns about the future of artificial intelligence and the idea that robots could eventually overtake humans are renowned physicist Stephen Hawking, Microsoft founder Bill Gates, and Elon Musk, the CEO and product architect of Tesla Motors and the CEO/CTO of Space Exploration Technologies (also known as SpaceX). While these leaders of technology see the potential of AI, they are cautious about how quickly these developments will occur and they share a common outlook that developments should be monitored and controlled.

An agricultural products conference in Villepinte, France, featured this weeding robot from the manufacturer Naio.

Not everyone shares these worries, though. Many engineers think that robots and advanced robotics will simply evolve the workforce. In New York City, automation has largely eliminated the job of elevator operator, yet society has survived. Future jobs may be centered on a more skilled workforce, and working class jobs will be the tasks robots take over more predominately. On the other hand, robots have the capacity to aid people immensely. IBM created Watson AI, which helps doctors search through research and other information to help diagnose patients. Will it take over a doctor's job? No. But Watson and other robots like it can help physicians do a better job of providing health care treatments and options.

CHAPTER FIVE

YOUR FUTURE IN ROBOTICS

Technology is ever evolving, and robotics will continue to grow and change. What does this mean for students looking into the future? It means opportunities for creative jobs, new career paths, and for ingenuity in engineering, computing, electronics, and more.

IN HIGH SCHOOL

For students showing interest in future careers in robotics, many high schools offer relevant courses and extracurricular activities. Coursework that young robot enthusiasts should look into include the usual STEM (science, technology, engineering, and mathematics) subjects. These may include calculus, precalculus, advanced algebra, computer science, physics, and

Robotics enthusiasts who start early learn pretty quickly that designing, engineering, and building robots often involves a team effort.

mechanical engineering, among others. Students with access to metal shop are encouraged to take these classes. Research to see if high schools focusing on technology careers exist near you. Such schools, which may require competitive application processes, will help prepare students for college-level work in these subjects, too.

COMPETITIONS AND CLUBS

There are many different clubs and organizations for students interested in robotics, and they are only growing in number. Young people can meet like-minded enthusiasts while sharing information and being mentored by older students, teachers, and club veterans in various aspects of robotics.

- **FIRST Robotics**. FIRST (For Inspiration and Recognition in Science and Technology) runs competitions for students in grades K through 12. Students work with advisers on a task given to them. Teams learn to design and build robots that are capable of completing these specific tasks. For high school students, this includes building a fully functioning robot that can compete in FIRST competitions.
- **FIRST Lego League**. For students in grades 4 through 8, FIRST Lego League teams research a real-world problem and develop a solution. Using Lego Mindstorms, teams must design, build, and program a robot that can compete on a table.
- **Botball**. Using STEM and writing skills, students design, build, and program a robot to compete in game competitions that change each year.
- **Project Lead the Way**. A national organization that develops STEM (science, technology, engineering, and

math) curriculum for K–12 classrooms. Students enrolled in PLTW learn technical skills as well as the ability to solve problems, think critically, and collaborate with others. PLTW encourages students to consider career paths in the STEM field.
- **BEST Robotics**. Short for Boosting Engineering, Science, and Technology, BEST holds a national six-week robotics competition for middle to high school students.
- **Carnegie Mellon Robotics Academy**. Carnegie Mellon's Robotics Academy studies how teachers use robots in classrooms to teach STEM subjects. It offers teacher training, curriculum development, and projects to help teachers provide cutting-edge robotics instruction.
- **iD Tech Camps**. Offered nationwide during summers, iD Tech Camps offer fun camps centered around STEM-related areas, including computing, coding, engineering, game design, and more.

COLLEGE MAJORS: CHOOSE YOUR FIELD

There are many majors to choose from, depending on the area of robotics that interests you most. Software engineering is a major in which students learn to design and write computer programs for computers, robots, smartphones, and other systems. Students take courses in computer science, software processing, software subsystems, advanced mathematics, including mathematical modeling, and more.

Computer science covers a wide range of topics. Instructors prepare pupils to work in intelligent systems (also known as artificial intelligence), computer graphics, computer theory, data management, distributed and parallel computing, systems software, and/or computer and information security. You might take

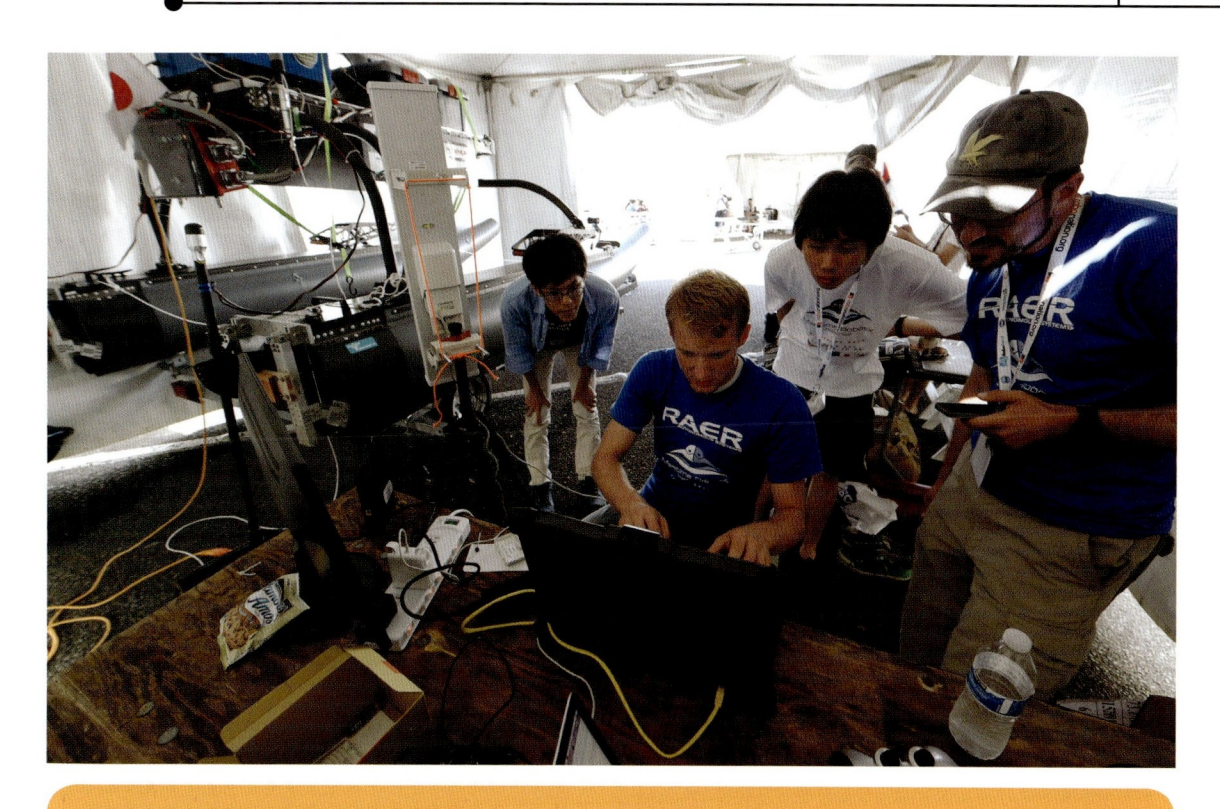

A student from Embry-Riddle Aeronautical University (*seated*) helps a team from Japan's Osaka University at the 2016 Maritime RobotX Challenge in Honolulu, Hawaii, a contest specifically geared to water-borne robots.

courses in computer science theory, programming, advanced mathematics, and software engineering.

Mechanical engineering encompasses the study of anything that has an engine or is mechanically based. This can include automobiles, planes, aerospace systems, energy-related technologies, and robotics. You might take courses in engineering mechanics, materials science, measurements, thermodynamics, fluid mechanics, design tools, physics, advanced mathematics, and more.

Manufacturing engineering technology is a field of engineering in which innovations in industrial productivity and technology

COMPANIES IN THE ROBOTIC REVOLUTION

Along with the many government institutions doing pioneering work in robotics, private companies have long been players in this business landscape. Some include:

- iRobot, the company that invented the Roomba, the self-propelled vacuuming robot, has now branched out to partner with leading universities in developing a robotic hand that is capable of grasping a pin and executing other dexterous motions.
- Google produces and uses driverless cars, and the company acquired Boston Dynamics, which makes robots that can walk, run, and jump.
- Touch Bionics has created a bionic hand that can be controlled with an app.
- Accuray invented the CyberKnife system, which employed robotics to accurately stream a beam of radiation to a tumor, which is safer and healthier for patients than the old method of radiation, which hit a large area around the tumor with radiation.
- Liquid Robotics produces robots that monitor sea life, marine vessels, ocean traffic, and more to keep our waters safer.

take center stage. You'll take courses that will prepare you to take on jobs in the areas of production systems design, mechanical development, and manufacturing productivity and efficiencies. Courses might include topics in manufacturing

The Swiss-based ABB Group, a major robotics and automation company, sponsors this training center in Berlin, Germany, as part of a larger, recent effort to educate 1,539 young technicians throughout the country.

processes, metals, strength of materials, circuits and electronics, automated control systems, robotics, automation, production processes, and supply chain management.

Electrical engineering encompasses the systems that transmit the signals from a robot's brain to its body. You'll complete courses in subjects such as semiconductor devices, embedded systems, electronics, mechatronics, electromagnetic fields, and transmission lines, plus advanced mathematics.

You can even study robotics as a minor or immersion topic. A minor or immersion is a small collection of courses that provide you with an area of expertise that supports or enhances your major. Many colleges, including but not limited to technically oriented ones, are adding minors in areas that provide targeted study in robots and robotics development. Minors in areas such as robotics, drones, data analytics, or artificial intelligence can provide a solid background in these areas and can help even those who may be planning to work with robotics in some way, even if they will not be building, designing, or programming robots.

THE FUTURE IS NOW

Robotics is a field that employs a range of professionals in areas as diverse as engineering, physics, computing, electronics, nanotechnology, artificial intelligence, sensors, software, and more. There are so many aspects that go into the creation and development of robots that career choices are almost endless for those who want to enter the field.

The possibilities remain endless, too, for the robotic creations that may dominate Earth's landscape—and the further reaches of space—as the current century progresses. Robotics could make it a very different world than we are used to, even if science fiction tales and current technology have somewhat prepared us. In many ways, our robot future is happening now, and it could lead us anywhere.

GLOSSARY

automation The engineering technique of making a machine that operates independently.

automaton A humanoid or animal-like object that moves on its own via simple engineering.

autonomous The ability to carry out tasks without outside control.

dexterous Skilled in movement, especially performed by the hands.

e-commerce Buying and selling products and services online.

evolution The slow but progressive development of something.

forecast To predict or calculate something in the future.

intuitive Something perceived based on intuition or feeling.

investment Financial support for something, including technological inventions.

mechanization The process of transforming workplaces from work done by hand to work performed by machines.

mechatronics A field that combines electronics and mechanical engineering.

precision Accuracy in performing a task.

productivity Describes how efficient a process or work performance is.

profit Money earned by a company or organization from sales of products or services.

programming Using computer languages to tell a computing system what to do and how.

repetitive Describing something that repeats or occurs over and over again.

sentient Describing an intelligence that is self-aware, such as that of a human.

vegetation Plant life found in a particular area.

FOR MORE INFORMATION

FIRST Robotics
200 Bedford Street
Manchester, NH 03101
(603) 666-3906
Website: https://www.firstinspires.org
Twitter: @FIRSTweets
FIRST (For Inspiration and Recognition of Science and Technology) Robotics inspires young people to be leaders in science and technology by engaging them in mentor-based programs that build science, engineering, and technology skills through team-based activities offered at elementary, middle, and high schools.

The Robotics Institute
Newell-Simon Hall
Carnegie Mellon University
Pittsburgh, PA 15213
(412) 268-3818
Website: http://ri.cmu.edu
The Robotics Institute, part of the of the School of Computer Science at Carnegie Mellon University, focuses on robotics becoming part of our everyday activities. Members explore the range of fields where robotics can have an impact, such as space robotics, medical robotics, industrial systems, computer vision, and artificial intelligence.

Western Canadian Robotics Society
c/o The Hangar Flight Museum
4629 McCall Way NE
Calgary, AB T2E 8A5
Canada

Website: http://www.robotgames.com
Facebook: @WCRS.YYC
The Western Canadian Robotics Society is dedicated to the advancement of "personal robotics" and organizing robotics competitions in western Canada.

Women in Technology
200 Little Falls Street, Suite 205
Falls Church, VA 22046
(703) 349-1044
Website: http://www.womenintechnology.org
Twitter: @WITWomen
Facebook: @WITWomenDC
The goals of WIT is to advance women in all areas of technology training and employment. The organization encourages networking and mentoring opportunities for women at all levels of their education and careers. WIT has student chapters at major universities, as well as local and national professional chapters.

WEBSITES

Because of the changing nature of internet links, Rosen Publishing has developed an online list of websites related to the subject of this book. This site is updated regularly. Please use this link to access the list:

http://www.rosenlinks.com/HOR/Future

FOR FURTHER READING

Brasch, Nicolas. *Robots of the Future* (Discovery Education: Technology). New York, NY: PowerKids Press, 2012.

Cassriel, Betsy. *Robot Builders!* (Scientists in Action). Broomall, PA: Mason Crest Publishers, 2015.

Ceceri, Kathy, and Sam Carbaugh. *Robotics: Discover the Science and Technology of the Future with 25 Projects* (Build It Yourself). White River Junction, VT: Nomad Press, 2012.

Clay, Kathryn. *Humanoid Robots: Running into the Future* (The World of Robots). North Mankato, MN: Capstone Press, 2014.

Hulick, Kathryn. *Careers in Robotics* (High-Tech Careers). San Diego, CA: Referencepoint Press, 2017.

Lacey, Saskia. *STEM Careers: Reinventing Robotics.* Huntington Beach, CA: Teacher Created Materials, 2017.

Nardo, Don. *How Robotics Is Changing Society* (Science, Technology and Society). San Diego, CA: Referencepoint Press, 2015.

Ryan, Peter K. *Powering Up a Career in Robotics* (Preparing for Tomorrow's Careers). New York, NY: Rosen Publishing, 2015.

Shea, Therese M. *The Robotics Club: Teaming Up to Build Robots.* New York, NY: Rosen Publishing, 2011.

Spilsbury, Louise A., and Richard Spilsbury. *Robotics* (Cutting-Edge Technology). New York, NY: Gareth Stevens Publishing, 2016

BIBLIOGRAPHY

Allonrobots.com. "Robot Surgery." Retrieved May 3, 2017. http://www.allonrobots.com/robot-surgery.html.

Atherton, Kelsey D. "These Magnetic Nanobots Could Carry Drugs into Your Brain." *Popular Science*, September 2013. http://www.popsci.com/technology/article/2013-09/minuscule-nanobots-carry-medicine-and-cells.

Atherton, Kelsey D. "US Army Wants Robot Medics to Carry Wounded Soldiers Out of Battle." *Popular Science*, September 24, 2015. http://www.popsci.com/army-wants-robot-medics.

Boyle, Rebecca. "Giant European Robofish Sniff Out Ocean Pollutants Autonomously." *Popular Science*, May 22, 2012. http://www.popsci.com/technology/article/2012-05/giant-robofish-sniff-out-ocean-pollutants-autonomously.

Boyle, Rebecca. "Hordes of Animal-Inspired Machines Lead to New Robotic Phylogenesis." *Popular Science*, July 8, 2011. http://www.popsci.com/technology/article/2011-07/hordes-animal-inspired-machines-lead-new-robotic-phylogenesis.

Boyle, Rebecca. "Rolling Robot Transforms into a Helicopter on Command, Decepticon-Style." *Popular Science*, May 16, 2011. http://www.popsci.com/technology/article/2011-05/rolling-robot-transforms-helicopter-command-decepticon-style.

Chalfant, Morgan. "Congress Told to Brace for 'Robotic Soldiers.'" *The Hill*, March 1, 2017. http://thehill.com/policy/cybersecurity/321825-congress-told-to-brace-for-robotic-soldiers.

Chang, Althea. "Pricy Robots 'Tug' Hospital Supplies." CNBC, April 30, 2015. http://www.cnbc.com/2015/04/30/pricy-robots-tug-hospital-supplies.html.

England, Rachel. "Robots Are Already Taking Over—We Interact with Them Every Day." Ecnmy.org. Retrieved May 4, 2017.

http://www.ecnmy.org/engage/robots-are-already-taking-over-we-interact-with-them-every-day.

Glaser, April. "11 Police Robots Patrolling Around the World." *Wired*, July 11, 2017. https://www.wired.com/2016/07/11-police-robots-patrolling-around-world.

Hessman, Travis. "The Dawn of the Smart Factory." *Industry Week*, February 14, 2013. http://www.industryweek.com/technology/dawn-smart-factory.

Jacobs, Rose. "Rise of Robot Factories Leading 'Fourth Industrial Revolution.'" *Newsweek*, March 27, 2015. http://www.newsweek.com/2015/03/27/rise-robot-factories-leading-fourth-industrial-revolution-311497.html.

Jung, Jaeyeon, and Atul Prakash. "Security Risks in the Age of Smart Homes." Conversation, May 2016. http://theconversation.com/security-risks-in-the-age-of-smart-homes-58756.

Krishnan, Armin. "Robots, Soldiers and Cyborgs: The Future of Warfare." Robohub, February 5, 2014. http://robohub.org/robots-soldiers-and-cyborgs-the-future-of-warfare.

Lafrance, Adrienne. "What Is a Robot?" *Atlantic*, March 22, 2016. https://www.theatlantic.com/technology/archive/2016/03/what-is-a-human/473166.

Miller, Claire Cain. "Evidence That Robots Are Winning the Race for American Jobs." *New York Times*, March 28, 2017. https://www.nytimes.com/2017/03/28/upshot/evidence-that-robots-are-winning-the-race-for-american-jobs.html?_r=0.

Newsweek. "Will AI Robots Turn Humans Into Pets?" April 14, 2017. http://www.newsweek.com/2017/04/14/will-ai-robots-turn-humans-pets-577691.html.

NPR. "Watch Robots Transform a California Hospital." May 27, 2015. http://www.npr.org/sections/money/2015/05/27/407737439/watch-robots-transform-a-california-hospital.

Parkin, Simon. "The Tiny Robots Revolutionizing Eye Surgery."

Technology Review, January 19, 2017. https://www.technologyreview.com/s/603289/the-tiny-robots-revolutionizing-eye-surgery.

Polland, Jennifer. "Watch: This Robot Is Poised to Change Surgery Forever." *Business Insider*, August 1, 2012. http://www.businessinsider.com/the-future-of-robotic-surgery-2012-7.

Popular Science. "Video: Rescuing Disaster Victims with Snake Robots Deployed by Dogs." January 7, 2012. http://www.popsci.com/technology/article/2012-01/video-dog-deployed-snakebots-are-future-disaster-rescue.

Robots.com. "Advantages and Disadvantages of Automating with Industrial Robots." Retrieved May 4, 2017. https://www.robots.com/blog/viewing/advantages-and-disadvantages-of-automating-with-industrial-robots.

Sofge, Erik. "The World's Top 10 Most Innovative Companies in Robotics." *Fast Company*, February 13, 2014. https://www.fastcompany.com/3026314/the-worlds-top-10-most-innovative-companies-in-robotics.

Violino, Bob. "The Future of Robotics: 10 Predictions for 2017 and Beyond." ZDNet.com. Retrieved May 4, 2017. http://www.zdnet.com/article/the-future-of-robotics.

INDEX

A

Accuray, 36
artificial intelligence, 4, 6, 9, 12, 16, 22, 29–31, 34, 38
automation, 13, 15, 17, 24, 27, 31, 37
automaton, 10
automobile production, 7, 13

B

BEST Robotics, 34
Botball, 33

C

Capek, Karel, 10
Carnegie Mellon Robotics Academy, 34
computers, 4, 9, 10–12, 28, 32, 34, 35
cost cutting, 13, 24, 25

D

Deep Blue, 11
Devol, George, 11
domestic robots, 15–17

E

education, 32–38
Engleberger, Joe, 11

F

FIRST Lego League, 33
FIRST Robotics, 33

Ford, Henry, 7–8, 10

G

Gates, Bill, 31
General Motors, 11
Google, 36
Greeks, 10
Grillbot, 16

H

hacking, 29
Hawking, Stephen, 30, 31
human error, 25

I

IBM, 11, 31
iD Tech Camps, 34
Industry 4.0, 24
iRobot, 36

J

job loss, 27

L

LEGO, 11, 31
Liquid Robotics, 36
Litter-Robot, 16

M

manufacturing, 4–5, 7, 13–15, 17, 22, 24–25, 27–28, 35, 36

Mars Rover, 20
mechatronics, 37
medical robots, 5, 17, 19, 29
military robots, 18–20
moon base, 22
Musk, Elon, 31
mythology, 10

N

nanorobots, 19
NASA, 20
national security, 4
Neato Botvac, 15

P

police robots, 6, 18–20, 26–27, 29
POS systems, 15
precision, 4, 7, 12, 17, 25
productivity, 24, 25–26
programs, 4, 5, 11–12, 14, 23, 29, 33, 35–38
Project Lead the Way, 33

R

reliability, 25
Robomow RS, 16
Roomba, 11, 36

S

safety, 15, 17–18, 25, 36
scientific robots, 20
service industry, 15
smart factories, 24
space, 4, 12, 20, 22–23, 31, 35, 38
supersoldiers, 19

surgical robots, 17, 19

T

Touch Bionics, 36
Tugs, 17
Turing, Alan, 10

V

vulnerabilities, 29

W

workforce, 21–31
workplace robots, 23–25

ABOUT THE AUTHOR

Laura La Bella is the author of more than forty nonfiction books for juvenile readers. Her interest in science, women's issues, and promoting STEM education has led her to write numerous books on these topics. Some of her titles include *The Goldilocks Zone: Conditions Necessary for Extraterrestrial Life*, *Drones and Law Enforcement*, and *Careers for Tech Girls in Gaming and Video Game Development*. La Bella lives in Rochester, New York, with her husband and two sons.

PHOTO CREDITS

Cover Erik Tham/Corbis Documentary/Getty Images; p. 5 Tomohiro Ohsumi/Bloomberg/Getty Images; p. 8 Library of Congress Prints and Photographs Division; p. 9 baranozdemir / E+/Getty Images; pp. 10, 19, 24, 30, 36 (background) Sylverarts Vectors/Shutterstock.com; p. 14 Andrey Rudakov /Bloomberg/Getty Images; p. 16 Wang Zhao/AFP/Getty Images; p. 18 Lucien Harriot/Getty Images; p. 21 Visual China Group /Getty Images; p. 22 Chip Somodevilla/Getty Images; p. 23 Chris Ratcliffe/Bloomberg/Getty Images; p. 26 Serge Mouraret /Corbis News/Getty Images; p. 28 Qilai Shen/Bloomberg/Getty Images; p. 31 Christophe Archambault/AFP/Getty Images; p. 32 DGLimages/iStock/Thinkstock; p. 35 U.S. Navy photo by John F. Williams; p. 37 Michele Tantussi/Getty Images.

Design: Michael Moy; Photo Research: Nicole DiMella

USN

Hassenfeld Library
University School of Nashville
2000 Edgehill Avenue
Nashville, TN 37212
www.usn.org